Feedbacks

From Causal Loop Diagrams to System Thinking Metodology

Juan Martín García

Feedbacks
Juan Martín García © 2024 Spain
info@atc-innova.com

Juan Martin Garcia is a renowned expert in System Dynamics and System Thinking, Ph.D. Industrial Engineer (Spain) and Postgraduate Diploma in Business Dynamics at the Sloan School of Management of the Massachusetts Institute of Technology (USA). He has extensive industrial experience in engineering and manufacturing, and 20 years' experience teaching building simulation models in large companies and universities, now teaches the online courses of Vensim in http://vensim.com/online-courses/. Some comments from their students are available in: http://atc-innova.com/comments.htm. He is the author of numerous books about System Dynamics published in English, Spanish, Portuguese and French.

CONTENT

Introduction

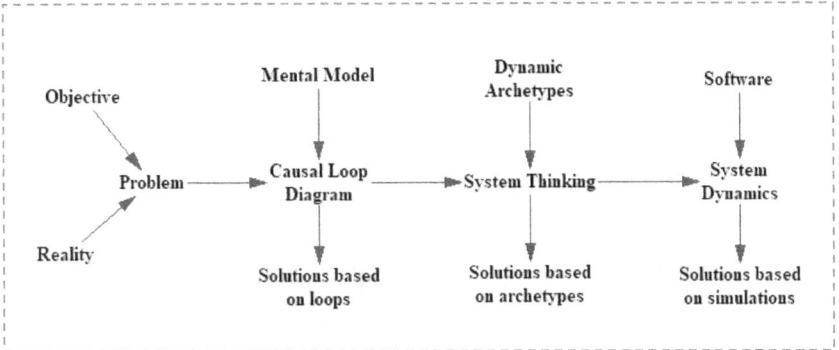

A general overview about problems and solutions

Introduction

We are increasingly aware that we live in a very complex and constantly changing reality and that it is more so every year. In order to make the decisions that are continually asked of us, we use mental models. However, these models don't always bring us closer to solving the problem, as the solution may be, as Jay Forrester calls it, *counter-intuitive*, even in the simplest of cases.

For instance, during a visit to the Science Museum with our children, we may have to explain why the hole in a water tank which is nearest the ground spouts water further than the spout above it. We may also have to explain why the image in a magnifying glass is inverted after a certain distance, instead of growing indefinitely.

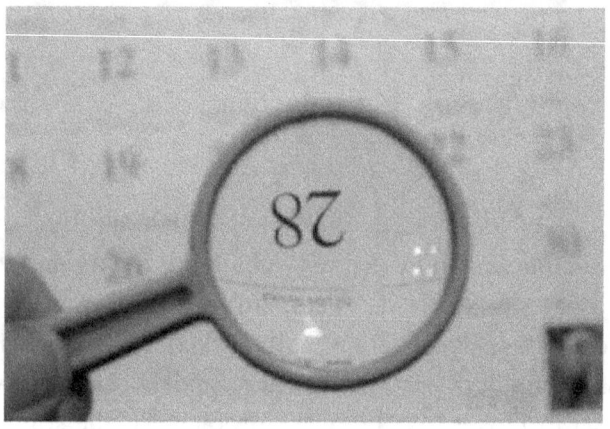

As Ludwig von Bertalanffy notes: for those who wish to study science, and only science, any posterior question makes no sense. *'Quod non est in formula non est in mundo'*. Such is the only legitimate position of science. Despite this, if we wish to further our understanding, there is only one analogy that can explain that which is irrelevant to the physicist, the analogy of the only reality we know directly - the reality of our immediate experience.

All interpretations of reality, to use Kant's expression, are an adventure of reason. There are therefore only two alternatives to choose from - either we reject all interpretations of the essence of things or, (if we do attempt an interpretation) we must remain conscious of its analogous nature, since we don't have the slightest proof that true reality is the same as that of our own internal experience.

When faced with the common occurrence of a reality with a limited number of parameters, especially when these are quantifiable, we employ formal models which allow us to act with a reasonably high probability of succeeding. However, when faced with complex situations with an uncertain number of parameters that are difficult to quantify, we can resort to less formal models that provide a more structured view of the problem, its key aspects, and possible solutions.

Lynda M. Applegate states that computers today are designed to treat information sequentially, instruction after instruction. This works well if the problem can be structured and divided into a series of stages. It doesn't work well with complicated, unstructured tasks which require intuition, creativity and discernment.

The main application of System Dynamics is in this kind of complex and loosely defined environments, where a human being's decisions tend to be guided by logic. We must remember that science is currently based on measurable and reproducible phenomena. As specialists in marketing know, people also behave according to certain rules which are fairly easy to measure and reproduce, for example, market law (more demand pushes up prices, etc.).

Computer models can provide information not attainable via mental models. They can show the dynamic consequences of interactions between components of a given system. When assessing the consequences of certain actions, the use of mental models means running the risk of obtaining erroneous conclusions. Intuition isn't reliable when the problem is complex. One possible reason for this is that we tend to think in terms of one-way cause-effect relationships, forgetting the structural feedback which almost certainly exists in such a system. When preparing a computer model, we must consider each step separately. The mental image we have of the system must be developed and expressed in a language that can be used to program the computer. Normally, any consistent and explicit mental image of any system can be expressed in this way. The mental images that we have of real systems are the result of experiences and observations. The explicit formulation of these experiences in a computer program forces us to examine, formalise and focus our mental images, thus providing us with a greater understanding through several perspectives.

Mathematical models, which are programmable, are explicitly expressed. The mathematical language used to describe the model leaves no room for ambiguity. A System Dynamics model is more explicit than a mental model and can therefore be expressed without ambiguity. The hypotheses upon which the model is built and the relationships between its constituent elements are present in complete clarity and are subject to discussion and revision. For this reason, a model's forecasts for the future can be studied in a completely precise way.

It is important to differentiate between the following two kinds of models: predictive models which are designed to offer precise information of the future status of the modelled system, as opposed to management models which are basically designed to decide whether option X is better than option Y. Management models don't require as much precision, since comparisons are equally useful. System Dynamics models are of the latter type.

As explained above, I understand the word *'system'* to mean a set of independent elements that interact with each other in a stable way. The first step towards understanding the behaviour of a system would be to define its constituent elements, and their possible interaction. The notion of Aristotle that the whole is more than the sum of its parts takes on a special meaning here.

The standpoint of System Dynamics is radically different to other existing techniques for the construction of socio-economic system models such as econometrics. Econometric techniques, which are based on behaviourism, use empirical data such as statistical calculus in order to determine the meaning and correlation between the various factors involved. The model is developed from the historical evolution of variables that are declared independent, and

9

statistics is applied in order to determine the parameters of the system of equations that link them to other independent variables. These techniques can establish the behaviour of the system without the need for information regarding its internal functioning. This is how stock market models analyse the upward and downward trends in the values of shares, the rising and falling cycles, etc. They are designed in order to minimise the risk of losses, etc. They don't attempt to gain any detailed knowledge of the internal workings of the firms, as the value of a given company rises and falls according to its new products, new competitors, etc.

The basic objective of System Dynamics is different. It aims to gain understanding of the structural causes of a system's behaviour. This implies increased knowledge of the role of each element of the system, in order to assess how different actions on different parts of the system accentuate or attenuate its behavioural tendencies.

One characteristic that sets it apart from other methods is that it doesn't aim to give a detailed forecast of the future. By using the model to study the system and test different policies, we will deepen our knowledge of the real world, assessing the consistency of our hypotheses and the effectiveness of each policy.

Another important characteristic is its long-term perspective, that is, that the period studied is long enough for all significant aspects of the system to evolve freely. Only with a sufficiently broad time scale can the fundamental behaviour of a system be observed. We mustn't forget that the results of certain policies are sometimes not the most appropriate, for example, if the time horizon of the decision-making process was too short, or if there was a lack of perspective when the problem was addressed. In these cases, it would be useful to know the long-term consequences of actions taken in the present, and this can be more tangibly attained if we use a suitable model.

The long-term development will be understood only if the main causes of any possible changes are identified. This process is facilitated if the appropriate variables are chosen. Ideally, the limits of the system should include the whole set of mechanisms that are responsible for any important alterations in the main system variables over a broad time horizon.

System Dynamics allows the construction of models after a careful analysis has been conducted of the elements of a system. This analysis allows the internal logic of the model to be extracted. Knowledge may then be gained of the long-term evolution of the system. It should be noted that the adjustment of the model according to historical data is of secondary importance, the analysis of the internal logic and the structural relationships within the model being the key issues involved in its construction.

(Note: All teaching material, and that includes this text, should be objective. This text aims to be so, but the author admits he hasn't always succeeded. For this he must apologise. Readers are invited to make their own assumptions as to what is an exposure of methodology, and what amounts to personal opinion.)

12

1. Identifying the Problem

What is the problem?

We are going to learn a method for constructing simulation models that help us determine the best solution for a given problem. These are therefore management models, not predictive models.

Firstly, we have to identify the problem clearly and give a precise description of the aims of the study. It may be obvious, but it is very important that the definition of the problem be correct, since all further steps depend on this. This is also very useful when establishing the amount of time and money that will be spent creating the model.

Once the core of the problem is defined, a description must be completed, based on the knowledge of experts on the subject, basic documentation, etc. The result of this phase should be a preliminary perception of the elements that have a bearing on the problem, the hypothetical relationships between them, and their historical behaviour.

The historical reference of a system is a record of the historical behaviour of the main elements that are believed to influence the problem. Where possible, they should be quantified. This is the graphical and numerical representation of the verbal description of the problem.

It's a good idea to ask ourselves whether it is necessary to construct a simulation model in order to find an efficient solution to the problem. This is an important question.

The construction of a model is a long and costly process. It can't be justified if there are other more simple ways of obtaining the same results. There are essentially two other ways - statistics and intuition.

- Statistics, or numerical calculus methods, are very useful for solving problems where there is an abundance of historical data or when we can assume reality will remain stable. For example, if you want to find out how many cars will drive past your house today, all you need is sufficient historical data and assuming the street hasn't changed, you'll get a good approximation.

- Intuition has got you where you are today, so don't underestimate it. For many problems, intuition provides the right answers, drawing on our experience and knowledge. Intuition is cheap and fast. Keep using it as often as possible.

Only when we can't apply one of these two options with certainty must we resort to constructing a simulation model.

Once the problem is defined, we will see that there are many directly or indirectly related aspects, or elements, which are also interrelated. They needn't be clearly or obviously interrelated. These elements constitute the system. We will now study reality as a system.

2. Defining the System

What is a system?

A system is a set of interrelated elements, where any change in any element affects the set as a whole. Only elements directly or indirectly related to the problem form the system under study here. In order to study a system, we must know the elements that make it up, and the relationships between them.

When we analyse a system we usually focus merely on the characteristics of its constituent elements. However, in order to understand the functioning of a complex system, we must focus also on the relationships that exist between the elements which form the system.

It is impossible to understand the essence of a symphony orchestra by merely observing the musicians and their instruments. It is the coordination that exists between them that produces beautiful music. The human body, a forest, a country, or the ecosystem of a coral reef are all examples of systems that are far more than the sum of their parts.

An ancient Sufi saying can illustrate this: *You can think, because you understand 'one', and you can understand 'two', which is 'one' plus 'one'. However, you must also understand 'plus'.* For example, in a traffic problem, many related elements converge - the number of inhabitants, the number of cars, the price of petrol, parking spaces, alternative transport, etc., and it is often easier and more effective to attempt to solve the relationships between the elements (*'plus'*) than the elements themselves.

A good method to begin defining a system is to write the main problem down in the middle of a blank page, surrounding it with the directly related elements. The elements which affect the main problem indirectly go around the appropriate direct elements. This will be the system that we will study in order to consider possible solutions to a given problem.

3. The Boundaries of a System

Where does the system end?

We have all heard the theory that a butterfly fluttering its wings in China could cause a tornado in the Caribbean. In our study, however, we will include only elements with a reasonable influence on the behaviour of a system. We mustn't lose sight of the objective, that is, to propose practical action towards effectively solving the problem at hand.

The system must contain as few elements as possible while providing a simulation that will truly allow us to decide which of the possible courses of action studied are the most effective solution to the problem. The models are generally small to begin with, with few elements. They are then expanded and perfected. Later on, elements which don't play a decisive part in the problem are eliminated. During the construction of a model, there are several extension and simplification phases in which elements are added and subtracted.

We can't ignore the relationship between the consumption of petrol and lung health. When we analyse the carbon combustion process in an electric power plant, we can see that, apart from energy, the following is produced - ash, suspended particles, SO_2, CO_2, etc. We can also see that there is no barrier between the desired product (electricity) and the by-products. Sometimes, the so-called *side-effects* are as real and as important as the *main effects*. The beauty of a system in nature is that the waste produced by one process serves to feed the next. Perhaps this is the model to follow for industrial design in the future.

The final size of the model must be such that its main aspects can be explained in ten minutes. Any model larger than this will fail.

4. The Causal Diagram

How do we represent the system?

The set of elements that bear relationship to the problem and that account for the observed behaviour, along with the relationships that exist between these constituent elements (which often involve feedback) form the system. The causal diagram represents the key elements of the system and the relationships between them.

As discussed above, it's important to draft versions that will bring us increasingly closer to the final complex model. The minimum set of elements and relationships that serves to reproduce the historical reference of a system is that which forms the basic structure of the system.

Once the variables of a system and the hypothetical relationships between these variables are known, we can move on to produce a graphical representation. This diagram shows the relationships as arrows between the variables. These arrows are marked with a sign (+ or -) which indicates the kind of influence one variable exerts over the other. A '+' means a change in the influencing variable will produce a change of the same direction in the target variable. A '-' means the effect will be the opposite.

So, when an increase in A results in an increase in B, or a fall in A causes a fall in B, this is a positive relationship, as shown below:

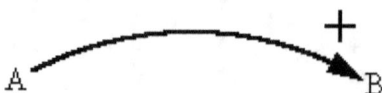

When an increase in A results in a fall in B, or a fall in A causes an increase in B, this is a negative relationship, which is expressed as follows:

5. Feedback

What is a loop?

A closed chain of relationships is called a loop, or a feedback loop. When we turn on the tap to fill a glass with water, the amount of water in the glass increases. The amount of water in the glass, however, also has an effect on the speed at which it is filled. We fill it more slowly when it is fuller. Therefore, a loop exists.

The system formed by us, the tap and the glass is a negative loop, because it is designed to achieve a goal (fill the glass without spilling). Negative loops act as stabilising elements in systems designed to reach a given goal, like when a thermostat in a heating system guides the temperature towards the level specified by the user.

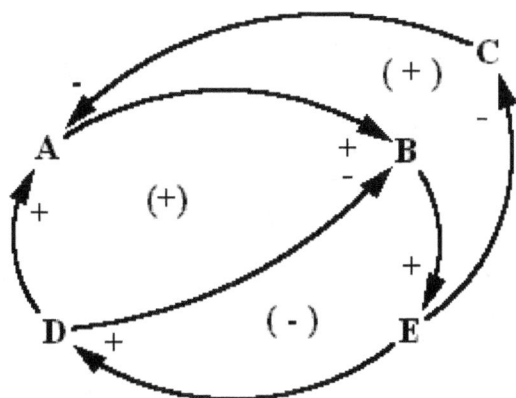

When we construct a model, loops appear. For example, those formed by ABEDA, DBED and ABECA in the following causal diagram.

Loops are defined as 'positive' when the number of negative relationships is even. If the number of negative relationships is odd, the loop is 'negative' (just as -3 multiplied by +3 gives -9).

Negative loops tend to stabilise the model, while positive loops tend to destabilise it, independently of the basic problem at hand.

Positive loops   Negative loops

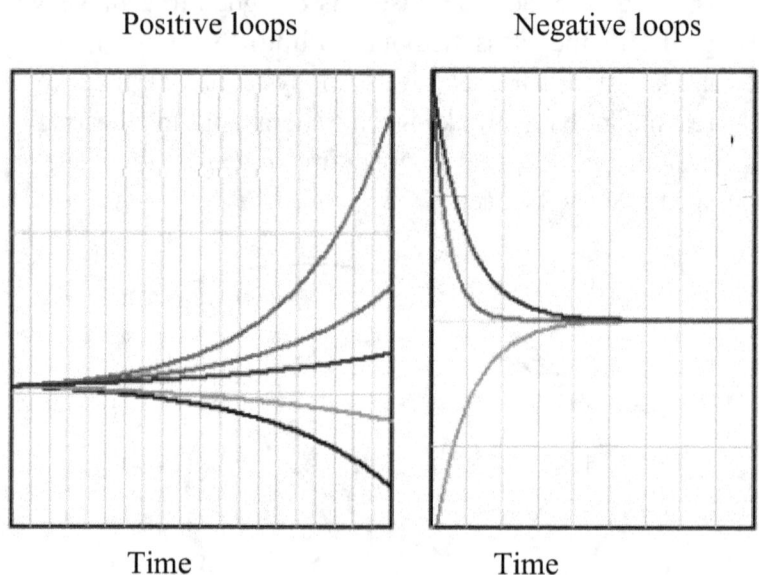

Time   Time

Real-life systems contain both types of loops, and the ultimate behaviour will depend on the dominant type at any particular moment. When a country acquires more arms it makes its neighbours feel threatened and causes them to also acquire more arms. This is a positive loop, also called a vicious circle, and grows more and more as it feeds itself. Positive loops cause growth, evolution, and also the collapse of systems.

Naturally, socio-economic and ecological systems are made up of hundreds of interconnected positive and negative loops and its ultimate behaviour isn't obvious.

The concept of the loop is very useful because it enables us to start from the structure of the system that we are analysing and work towards its dynamic behaviour. If a system fluctuates persistently, remains in equilibrium or drops off rapidly, we can identify the structural reasons and decide how to go about modifying the causal loops that are going to influence it. This procedure can be applied to anything from the control of an industrial process to the monitoring of diabetes or cancer, fluctuations in the price of raw materials, or economic growth.

Yet the most important use of this concept is in understanding **how the structure of systems affects their behaviour.** In the same market and in the same year, various firms that offer the same product present very different economic results. The less competent managers put this down to causes beyond their control – the cost of labour, competitors, customers' habits, and so on – when in fact they should study why the systems they control (their businesses) have a less competitive structure than those that show better results.

Country A perceives that the arms race was caused by country B and vice versa. But in reality we can also say that country A has caused its own rearmament by acquiring arms, as this causes the rearmament of country B. Similarly, the rise in oil prices is due both to the concentration of production in a few countries and to excessive consumption in developed countries of a product that's limited, inasmuch as it isn't renewable.

Identifying the cause of a problem as being something that is not external to the system doesn't tend to be very popular as it is easier to blame external factors beyond our control. The trouble is, if the exponents of the argument of the external cause really believe what they are saying, they will be unable to identify the true cause of the problem – inside the system – and obtain the desired results. If the system contains the elements that cause the problem, it also has those that can be used to solve it.

For example, a product's life curve can be said to be regulated initially by a positive loop that permits rapid exponential growth, followed by a steady state dominated by a negative loop involving the saturation of the market, and finally a usually sudden drop caused by the appearance of fast-growing substitute products.

Lastly, note that the causal diagram is very important for explaining the final model to the user if he or she isn't familiar with this technique, as is usually the case.

6. The Limiting Factor

The limiting factor is the element of the system that is limiting the growth of the system at this moment in time. There's only one limiting factor at any given moment, although over a period, various different elements of the system can act as limiting factors.

Maize can't grow without phosphates no matter how much nitrogen we add to the soil. Although this fact is quite elementary, it is often ignored. Agronomists assume that they know how the soil should be fertilised as they know the 20 main substances that plants require for nutrition, but how many elements are they unaware of? Attention is often focused on the more voluminous substances but seldom on the truly important parameter - the limiting factor.

In order to understand reality we have to appreciate not only that the limiting factor is essential but furthermore, changes also modify the elements that make up the system. The relationship between a growing plant and the soil, or between economic growth and the resources that sustain it, is dynamic and constantly changing. When a factor ceases to be limiting, growth occurs and the proportion between the factors changes until one of them becomes the limiting factor. If we can direct our attention towards the next limiting factor we can advance towards real understanding and efficient control over the evolution of systems.

The limiting factor is dynamic; in the growth of a plant, today it might be lack of water, whereas tomorrow this might be solved and the limiting factor might be lack of nutrients, and so on. There is never more than one limiting factor in a specific moment.

7. The Key Factors

Key Factors are also called leverage points, as it is here that pressure or influence is exerted. **A system includes several key factors, and they tend not to vary over time.** We can use they bring about major changes in the system with minimum effort. They can unleash violent behaviour in the system. Each system has a number of key factors and they are neither obvious nor easy to identify.

A normal person's key factors will be related to their health, the family and (hopefully) their education. **They are the driving force behind their acts in their daily lives.**

We also have to take into account that these **key factors can unleash violent behaviour.** Sometimes people tolerate all kinds of humiliation, publicly and privately, yet a derogatory remark about their parents can be fatal. This is, then, a key factor. Key factors can be physical (we can stick a finger in somebody's ear without making them unduly angry, but not in their eye) or psychological (a minor accident in a new car can make some people react extremely violently).

In order to attain a goal, huge efforts are sometimes made in the wrong directions. This is especially true in the personal, social, business and ecological fields. In an attempt to avoid this, Jay Forrester proposed a set of guidelines for the business world that can easily be extrapolated to other areas.

1) Whatever the problem is that has arisen, it is necessary to know the inner workings of the

system, how it takes its decisions, how it operates. Don't be led astray by indications that point towards momentary or superficial factors, however visible they may be.

2) Often a small change in one or very few policies can solve the problem easily and definitively.

3) The key factors tend to be ruled out or judged to be unrelated to the problem at hand. They are rarely an object of attention or discussion and when they're identified, nobody can believe that they are related to the problem.

4) If somebody happens to have already identified a key factor, it is not unusual for action to have been taken in the wrong direction, thus seriously magnifying the problem.

Models enable us to conduct sensitivity studies and see which of the system's elements can have a decisive bearing on its behaviour. In other words, they enable us to identify the key factors. However, that doesn't mean we can't advance without their help.

The peculiarity of these key factors is that they are located in unexpected points or aspects that provoke counter-productive actions. This is difficult to illustrate with a causal diagram. The phenomenon seems to be attributable to the difficulty in interpreting the behaviour of a system that is already defined, rather than to any specific structure, as the effect of the interrelationships is beyond our capacity for analysis. (For me, this means that the system has more than three loops).

This inability to perceive and interpret the nature of the system and the identity of its key factors make for counter-intuitive behaviour by the system, with the result that our actions are in the wrong direction. Let us take a look at some examples.

a) A car engine manufacturing firm suffered a constant loss of market share. Every four years there was a major loss of customers who seldom came back. According to the firm's analyses, the problem lay in their policy on stocks of finished products. The company was reluctant, due to the high financial cost, to keep a large number of engines in stock waiting for orders to arrive. The policy was to keep stocks of finished products low. This policy saved a great deal of money but whenever there was an upturn in the economic cycle, the firm was overwhelmed with orders that they were forced to attend to with long delays. The customers then went to the competitors who supplied the engines more quickly. The firm responded to the loss of sales with a programme of cost-cutting measures, including further reductions in stocks of finished products.

b) Dairy farms are steadily disappearing. Measures are proposed to combat this, including tax cuts, soft loans and subsidies. There is plenty of incentive for anybody wanting to start up a small farm. However, the main reason why farms close is expansion. Farmers try to increase their income by producing more milk. When all the farmers do this, the same the market is flooded with milk and prices fall (as there is no intervention or guaranteed price. If there were, the burden would be shifted to the external factor). When the prices have dropped, each farmer has to produce more milk in order to maintain his or her earnings. Some manage to do so and others don't, and of the latter, those that are in the weakest position give up farming.

c) One of the key factors in any economy is the useful life of the installed capital. The best way to encourage the sustained growth of the economy is to stretch this useful life as long as possible. Yet the policy that is practised is one of accelerated obsolescence, or priority is given to replacing existing equipment with machinery designed to provide short-term economic growth.

d) The right way to revitalise the economy of a city and ease the problem of depressed areas occupied by people without economic resources isn't to build more subsidised housing. The solution is to demolish the abandoned factories and houses and create space to set up new businesses, thus allowing the balance between jobs and population to restore itself.

Ideally, we'd have a set of simple rules to find the key factors and know in which direction to act. It is not always possible to find these points by simply observing the system and this is where computer simulation models really come into their own.

8. Classification of Systems

8.1. Stable and Unstable Systems

A system is stable when it consists of or is dominated by a negative loop, and is unstable when the loop is positive. That is, when the dominant loop contains an odd number of negative relationships, we have a negative loop, and the system will be stable. The basic structure of stable systems is as follows:

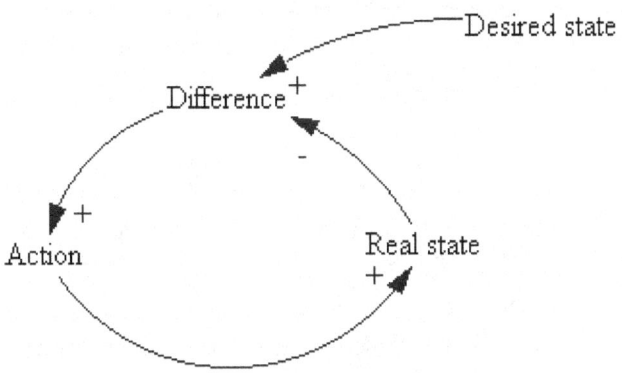

Here we can see that the system has a 'desired state' and a 'real state'. These two states are compared ('difference'), and on the basis of this value, the system takes 'action' to move the 'real state' towards the 'desired state'.

31

In this case the initial parameters are of relatively little importance, since the system will act according to the environmental conditions it encounters, so if it is hungry, it will look for food and when it finds it, it will deal with its next objective, and so on.

It is important to note that in stable systems the structure that generates the behaviour is always the same, that is, there is an odd number of negative relationships and the loop is negative.

This means that the system permanently compares its real state with the desired state and when there is a difference, it takes action to bring its real state closer to the desired one. Once these two states coincide, any change in the real state will result in action (proportional to the difference) to regain the desired state.

This is how we usually find systems. By the time we get close to them, they are in a position of stability. If a system is unstable, we are unlikely to be able to study it as it will have disintegrated before we can analyse it. However, if we are designing a totally new system, we should take the trouble to find out whether it is going to be stable and if we are designing a change in a stable system, we have to ensure that we are not changing it into an unstable one.

Examples of systems that are not in an optimum situation but carry on over the years – i.e., stable systems – can be found in many fields, for example, government, workers and bosses together produce inflation that is harmful to all. Rich countries and poor countries trade with raw materials, each with a different political and economic objective, and the result is permanent price instability.

Let us suppose that the government intervenes in the system with a particular policy that puts the state of the system where it wants it. This will cause major discrepancies between the other elements of the system which will intensify their efforts until, if they succeed, the system is back very close to the initial position after each element has made a huge effort. For example, I think in the work that has gone into improving the traffic in the city Barcelona over the last 10 years. The traffic improved for a few years after the opening of the Ring Roads, but now we are faced with the same problems as before – except that they affect many more cars.

The most effective way of combating the natural resistance of the system is to persuade each element to change its objectives in the direction in which we want to lead the system. Then the efforts of all the elements will be directed towards the same goal and the effort will be minimum for all as they won't have to resist the tide going the other way. When this can be achieved the results are spectacular. The most common examples of this are the mobilisation of the economy in wartime and the recovery after wars or natural disasters.

A less warlike example can be found in the birth rate policy in Sweden in the 1930s, when the birth rate fell below the rate of natural replacement. The government made a careful assessment of its objectives and those of the population and found that an agreement could be reached on the basis of the principle that the important thing isn't the size of the population but its quality. Every child should be wanted and loved, preferably in a strong, stable family, and have access to excellent education and health care. The Swedish government and citizens agreed on this philosophy. The policies that were introduced included contraception and abortion, education on sex and the family, unhindered divorce, free gynaecological care, aid for families with children in the form of toys, clothes, etc., rather than cash, and increased spending on education and health. Some of these policies seemed strange in a country with such a low birth rate, yet they were introduced, and since then the birth rate has risen, fallen and risen again.

Some systems lack feedback and the models we build must show that fact. For instance, if we know the initial parameters of a clam (type, weight, etc.) and we control the environmental conditions in which it will live, we can safely predict its weight after 6 months. There's a 'transfer function' between the start and end values and we have to find it, but that's all. Other examples: God is someone who gets his real state to coincide with his desired state instantly. Suicide is the response of those who perceive that they will never get their real state to coincide with their desired state and that therefore, all action is pointless. Please note: The more intelligent a system is (i.e., the clearer its vision of its objectives), the more stable it will be. This is applicable to people.

8.2. Hyperstable Systems

When a system consists of several negative loops, any action taken to modify one of its elements is offset not only by the loop in which that element is located but also by the whole set of negative loops which act to support it, thus superstabilising the system.

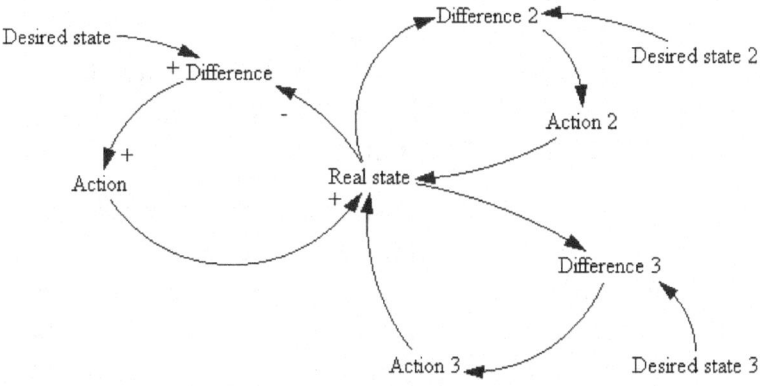

An analysis of the system can be helpful. Any complex system, whether social or ecological, is made up of hundreds of elements. Each element is only linked to a limited number of variables which are important to it, and which it permanently compares with its objectives. If there is a discrepancy between the state of these variables and its objectives, the element acts in a particular way to modify the system. The greater the discrepancy, the more energetic the action taken by the element on the system. The combined action of all the elements that attempt to fit the system to their objectives leads the system to a position that none of the elements actually wants but in which all of them find the smallest gap between the parameters that are meaningful for them on the one hand and their objectives on the other.

Why do many problems persist despite continual efforts to solve them? As we have just seen, systems base their stability on the actions of all its elements in pursuit of different objectives, trying to get the rest of the system as close as possible to its desired position. From this moment on, if an element of the system or an external agent attempt to modify its stability, the other elements will take action to go back to the initial situation, thus neutralising the action that altered its stability.

So the answer is simple - systems resist any change we try to introduce because its present configuration is the result of many previous attempts like ours (unsuccessful ones, otherwise the system would be different today) and an internal structure that renders it stable and capable of neutralising changes in its surroundings, such as the one we made with our action. The system achieves this as a whole, by rapidly adjusting the internal relationships between its elements in such a way that each continues to pursue its own goal and together they neutralise the action exerted on them from outside.

8.3. Oscillating Systems

We will see later in the case studies that for a system to display oscillating behaviour it has to have at least two stocks, which are elements of the system that produce accumulations.

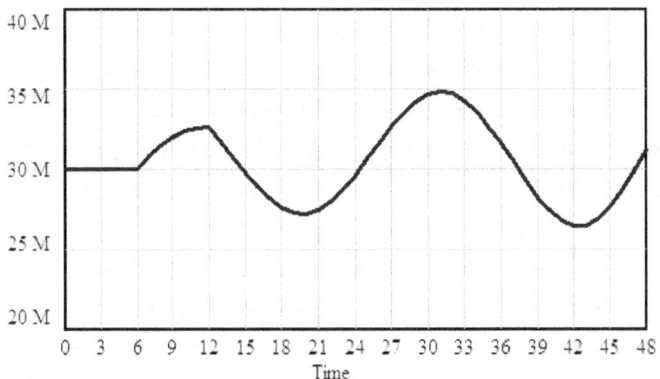

8.4. Sigmoidal Systems

These are systems containing a positive loop that acts as the dominant feature at the beginning, causing the system to undergo an exponential take-off. Subsequently, control of the system is taken over by a negative loop that cancels out the effects of the

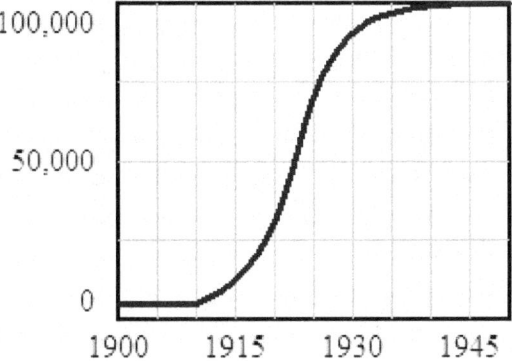

earlier positive one and provides the system with stability, setting it to a particular value asymptotically.

It is important to keep sight of the fact that in this case, we are dealing with the same system all the time, dominated by one part of it in one period, and by a different part later on. So in order to regulate its behaviour, we will have to find a way to play up or down the part of the system we are interested in. We also have to be aware that in the mid-term the negative loop will stabilise the system at its target value. All we can do is regulate the time scale and the way in which the system reaches its objective.

9. Generic Structures

In complex systems, we can observe the same structure, that is, desired state - real state - difference - action, over and over again in very different contexts. On top of this base structure, generic structures have been identified that tend to appear regardless of the object of study.

There is always the same 'intelligent' structure that seeks to bring the real state closer to the desired state.

9.1. Resistance to Change

When new managers joins a firm, usually with new objectives, they often find that its employees put up resistance to everything they propose – 'They have already tried that. That won't work here. Our customers like it the way it's always been. That proposal is very risky.' In short, the company acts as a system that has managed to survive innumerable economic crises in the past, and as a structure, is capable of neutralising any change, whether from inside or outside, due to the multiple relationships between its members. Each pursues a different objective, yet as a whole they have succeeded in endowing the firm with stability, although that doesn't mean its position is necessarily the most efficient. For this reason it is often wise for new managers to seek the commitment of the general manager for their new objectives as a way of achieving a certain amount of strength and aligning the other elements in the company towards these objectives.

Many systems are not only resistant to new policies designed to improve their state (greater productivity, lower costs, etc.) but also show a persistent tendency to worsen it, despite the efforts to improve the situation. Examples abound in the business world - productivity, market share, quality of service, etc. And on a personal level, we all know somebody with a tendency towards obesity in spite of repeated diets!

9.2. Erosion of Objectives

The action required to shift the real state towards the desired state always demands an effort and this effort in turn requires a consumption of time, energy, money, etc.

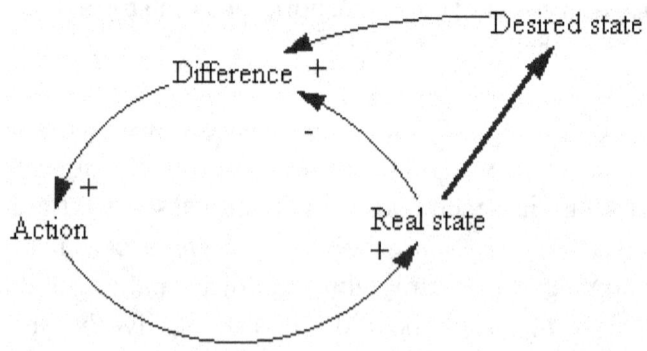

It is normal for the real state to 'contaminate' the desired state, that is, **for the system to try and**

avoid the consumption of energy required to take the action. The desired state is initially reconsidered, since if it coincided with the real state, no action would be necessary. The diagram below shows this 'contamination'.

If contamination occurs, the desired state is modified until it is the same as the real state. The difference is then zero and therefore there is no action to be taken. And so the real state of the system doesn't change.

There are only two ways of avoiding this process:

1.- Find a 'hero' system. That is, convince the system that it doesn't matter how much effort is required to reach the desired state, it just has to be reached. (Personally, I can assure the student that this way of avoiding the contamination process doesn't tend to get results in the 21st century.)

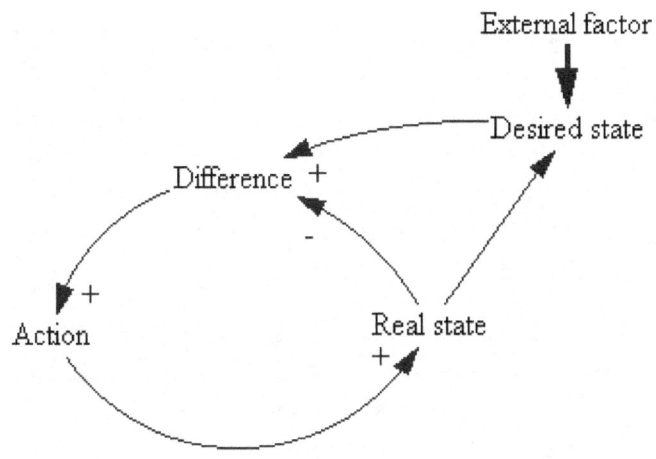

2.- Get an 'external element' to serve as a reference or anchor for the desired state, so that it can't be altered by pressure exerted on the system, and so that the system has no capacity to alter the 'external element'.

In Spain, when secondary school students consider the possibility of carrying on studying at a public university, they already have a fairly accurate idea of the minimum mark they have to get in their schoolwork and the entrance exams. Their desired state is that minimum entrance mark. It's not negotiable. Their real state tends to be a lower mark than the entrance mark in the early years of secondary education, so they perceive a difference, which leads them to take action (studying harder) in order to get their real state to match the desired state. If students know what it is they want to study, their family doesn't need to push them at all. The system isn't contaminated because the desired state (the minimum entrance mark) isn't alterable.

Later, when they start at university, if you ask, they'll say they want to be a great professional and that they're going to get an average mark of 10 in their degree. With the first exams come the first failures, which make them: 1) study harder than they'd anticipated, and 2) tone down (contaminate) their desired state, from the desired 10 to the non-negotiable minimum of 5.

The structure that brings about this behaviour is based on the idea that the system includes a particular objective (e.g., desired weight) that is compared with reality (real weight), and the discrepancy between these two values triggers an action which is proportional to the size of the gap. This is the usual pattern, seen up to now as a negative loop that tends to gradually pull the system towards its objective if it encounters some discrepancy. However, sometimes the state of the system can condition or modify the desired state, either because the real state is very long-lasting, because the action taken involves a great effort, or indeed for some other reason. The initial goal shifts towards the real state of the system.

This relieves the need to take action as the discrepancy has been reduced, not because the system has approached the objective but because the objective has approached the real state. As a result, the action taken is smaller.

In the case of the weight of obese people, this occurs when they accept that the target weight was too ambitious and that a more realistic target (a higher weight) is better. This argument serves as an excuse to follow a less strict diet. When they see their weight doesn't fall, they reconsider the ultimate target once again, and so on, until they think that actually their real weight is best, at which point they don't have to follow any sort of diet (this would have involved a sacrifice).

There are plenty of examples of this pathology in environmental pollution, law and order, traffic accidents, etc. In all of them, a poor performance becomes the standard in the face of the effort required to do something effective.

A system that bases its objectives on reality and intends no more than to improve on it is permanently drawn towards poor results. A system that gets its targets from outside itself is immune to this type of process.

It may seem paradoxical but if a student is convinced that he must pass all his subjects in July because his father has imposed it as an immovable objective, for whatever family reasons, it will be easier for him than if he himself had made that decision. If it's a personal decision it can be reconsidered when some of the subjects prove to be too difficult. He can accept to leave one or two for next time, which means less studying. However, if the objective is non-negotiable, this risk doesn't arise, and he has to study as hard as necessary to reach the objective.

Economics provide any number of examples. In Spain, nobody remembers such low rates of inflation as there are now. Any government would be satisfied and would be happy to give up reducing inflation further, as that would mean taking very unpopular measures (a wage freeze for civil servants). If the target for inflation was in the hands of the government, corrective measures would have been less strict in the past and the present, since they would have meant less public spending and therefore lost votes. However, the target for inflation was imposed as a condition for entering the euro zone and as such was beyond the control of the government which pulled out all the stops and took all the unpopular measures they deemed necessary because there was a fixed goal with a deadline and it was non-negotiable.

The obvious antidote to this pathology is to fix absolute objectives for the system that are not based either on the past or the present situation and take corrective measures depending on the difference.

An absolute objective loses credibility if it is raised or lowered and it won't get it back. We see this sometimes when an objective is raised because the initial objective has been reached. When this happens, everybody expects the initial objective to be changed again (but this time downwards) when the results are lower than the initial objective.

9.3. Addiction

Sometimes the real state of the system matches the desired state not as the result of action but due to support from outside the system. This support may or may not be permanent and may or may not be of interest, but the net effect is to bring the real state into line with the desired state, resulting in zero difference and therefore action by the system is unnecessary.

This phenomenon occurs when there is an objective that serves as a point of comparison with the state of the system. On the basis of the discrepancy observed, corrective measures are taken proportionally but in this case, the action taken doesn't serve to bring the system's real state closer to its desired one but rather to create the perception that the real system is close to the desired one, whereas in fact this action has no such effect.

The lack of clear perception of the real system leads to a situation in which the necessary corrective measures aren't taken because the state of the system is perceived as being closer to the objective than it really is.

When the immediate or short-term effect of the action disappears, the problem (i.e., the discrepancy between the real state and the desired one) reappears, often with greater intensity, so the system reapplies some measure that appears to solve the problem whenever the effect of the previous measure starts to fade.

Alcohol, nicotine and caffeine are obvious examples of addictive substances. Another case that springs to mind is the use of pesticides, which eliminate, together with the pest in question, the natural control mechanisms. As a result, the pest will reappear as soon as the effect of the pesticide abates, but this time without any natural control.

In cases of addictive systems, it is difficult to find suitable policies, since the action taken offers apparent results in the short term, but once the process is rolling, it is difficult to stop. Obviously, the best approach is to be aware of these types of processes, in other words, to be wary of using measures that attack the symptoms but make the system worse when they are relaxed. Once the addictive process has been started, you have to expect at least short-term difficulties if you plan to stop this process, be it physical pain for somebody who takes an addictive drug, rising petrol prices on inclusion of the associated environmental costs, or more pests and lower-quality food until such time as natural predators return.

Sometimes it is advisable to wean yourself off an addiction gradually. But it is always less costly to avoid the addictive process in the first place than to stop it later.

9.4. Shifting the Burden to the External Factor

As they get older and spend more and more time reading, some people gradually get poorer eyesight. In the end they can't read what is written on a blackboard, and can't renew their driving licence, so they get glasses or contact lenses. Then, in one year their eyesight worsens as much as it did in the previous 30 years. So their glasses become a necessity not only to see at a distance but also to read a document. Apparently this happens because for years, the muscles around the eyes have been straining to compensate their poor vision and when this effort is no longer necessary, they cease to act and end up losing this ability totally. Before long, they need stronger lenses.

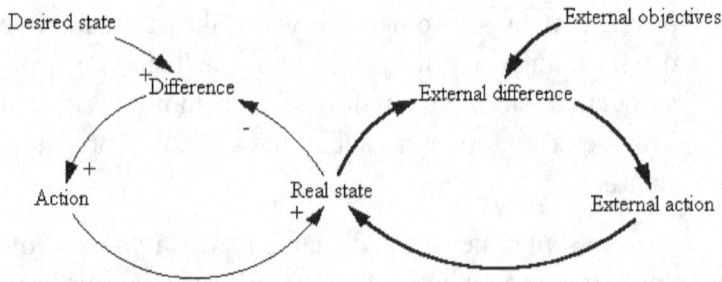

This is a classic example of shifting the burden to the external factor. In this sort of system an external force keeps the system in the desired situation. A well-intentioned, benevolent and very effective force decides to help us to get the system where we want it. This new mechanism works very well.

But with this process, through the active destruction of the impediments that redirected the system towards the desired position, or simply through atrophy, the original forces that worked to correct the position of the system are weakened. When the system moves away from the desired position the external factor makes an extra effort, which weakens the original forces still further. In the end the original system adopts a position of total dependence on the external factor, as its original corrective forces have disappeared completely and in most cases, irreversibly.

It is easy and fun to find other examples of shifting the burden to the external factor. Here's the start of a possible list.

Problem	System	External factor
Difficulty of calculating	Mental skill	Calculators
Care of the elderly	Family	Social Security
Lack of pasture in winter	Wild deer	Provision of fodder
Infections	Human body	Antibiotics

Seeking the aid of an external factor to get the system where we want it to be isn't in itself a bad thing. Usually, it is beneficial and enables the system to better tackle objectives. Yet the dynamics of the system can be problematical for two reasons. Firstly, the external factor that intervenes doesn't tend to perceive the consequences of its help on the elements of the system, particularly on those that performed the same task as itself. Secondly, the community that has helped today doesn't stop to think that this help is temporary; they lose their long-term perspective and so become more vulnerable and dependent on the external factor.

The withdrawal of aid from a system that is being helped, whether it is the human body, a particular area of ecological value or a human community, doesn't tend to be easy and is often simply impossible. This process of withdrawing help without harming the system must be based on identifying the internal elements of the system that in their original state, took care of correcting the problem, strengthening these mechanisms and, as they begin to do their job, gradually withdrawing the help.

9.5. Short and Long-Term Effects

A rational analysis of the problem at hand based on our capacity for synthesis and our ability to imagine things seems to be a bad guide to finding the key factors. We generally pay attention to the components of the system and their behaviour in the short term, all on the basis of incomplete information. Consequently, firms reduce their stocks of finished products when sales are seen to slump, the government extends its tax reductions for small farmers, and policies are introduced to encourage firms to replace their machinery instead of maintaining it properly.

They are all very reasonable policies. But there's still something inside us that just might make us realise that our customers' dissatisfaction with our long delivery schedules, or farmers' permanent concern with increasing their output, or the idea of replacing a machine that's productive, all means something, but we haven't given it the right interpretation.

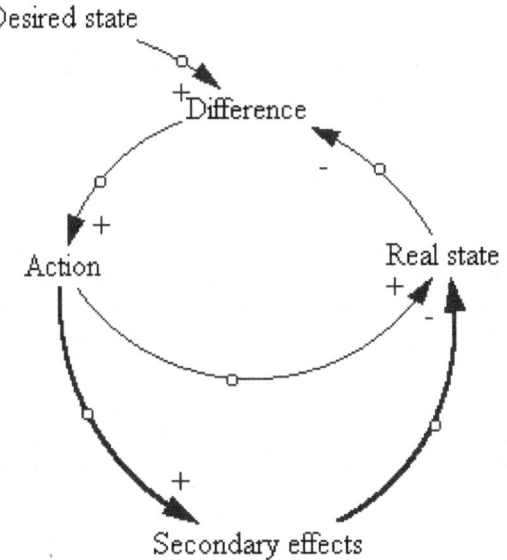

Desired state

+ Difference

Action

+

-

Real state

+

-

+

Secondary effects

Finally, I'd like to say that in my opinion, we have the capacity to understand not only simple systems but complex ones too, and to find the key factors. What we don't appear to have is the capacity to articulate the arguments to convince others or even ourselves that what we're perceiving is right. We expect the solution to be closely related to the symptom; we expect long-term profit to start with short-term profit, or a strategy that's satisfactory for all the agents involved.

Yet we know complex systems don't behave that way. So something inside us still insists somehow that maybe that simple, effective solution isn't the best. And then we carry on proposing policies that can't work, denying ourselves other simpler and more effective ones that could.

We try to compete instead of cooperating, and we try to reach the limits of the environment's capacity instead of admitting that we have already gone too far. The results are famine, war, pollution and depression. And right in front of us, within reach of our capacity for understanding, stand balance between countries, peace, equality and sustainable development.

10. World Models

For most people, especially engineers, world models conjure up enormous computers with huge amounts of information about any topic you can imagine. The first well-documented world simulation model was relatively simple. It was published in 1971 by Jay Forrester. Since then, the proliferation of world models has been massive. These models have been created in different parts of the world, using different techniques with very diverse goals in mind. Even so, making these models is clearly limited to the amount of information their designers can include. Each of these models uses just a minute amount of the information that is available about the world. Most of them focus on economic, demographic and agricultural production factors. A small number deals with environmental problems and the use of natural resources. Almost none refer to war, politics, new ideas or natural disasters. Most assume technology doesn't change or changes automatically and exponentially without a price tag, allowing more and more production at an ever decreasing cost. Some models show the world as a simple unit and a continuous whole, while others divide it into 10 or 15 regions, and still others divide it into a hundred nations. Some use a time span to the year 2100 and others just project a few years into the future. Some, especially the former, raise heated discussions and others have been explicitly instructed to refute the conclusions of previous ones.

We are going to see some of the characteristics of these world models. They are frequently misunderstood or interpreted by an audience that is either quite gullible or quite sceptical about computer models.

1. There is quite an assortment of models. They have been made by people with defined political and cultural beliefs and they, therefore, usually tend to be quite slanted in different ways. There is no such thing as a model in the socio-economic, ecological or other realm which is 'objective'; in other words, neutral, and therefore, not influenced by those who have made it.

2. The models, especially those of the world, are tremendously complicated in what they refer to (detailed structures by age of the population, varied economic sectors, complex sketches about business and extensive classification of people by income) and amazingly simplistic in what they omit (weapons, age of infrastructures, motives, social norms, political structures, the starting point and destination of the flow of raw material, etc.).

3. A model is a list of mathematical equations that explicitly sums up a view of the world. It is bolstered by statistical parameters and logical consistency that is able to produce phrases such as: 'If all of these hypotheses are correct and there aren't any others which should be considered and they are still valid in the future, then the logical results are...' (The reader of this text can copy this just in case they need it some day).

The importance of these models perhaps resides not so much in them but in the fact that underlying them is the effort of people from various continents and ideologies interpreting the world from different points of view, given the limitations of traditional work tools. They have all had to observe the world as a whole and pondered the long-term implications of the ties that connect population, capital and the production of goods and services that link all nations. They have immersed themselves in global statistics and constructed a model that has portrayed the global situation generally and relatively consistently - each buyer needs a salesperson; each birth must eventually be linked to a death. Once capital is set, its original use for a tractor factory cannot be changed for a hospital.

Consequently, although some topics and specific details are given a different perspective, there are some common conclusions and sensations obtained when observing the world as a closed system in all the models. The model creators, who are generally hostile and critical toward each other, are surprised that their basic conclusions coincide.

The following points sum up these basic conclusions:

1. There is no physical or technical reason why physical or technical needs cannot be met in a reasonably predictable future. These needs are not met now because of political and social structures, as well as values, norms and world views, not because of physical limitations.
2. The population and consumption of physical resources cannot grow indefinitely on Earth.
3. Clearly and simply, information is incomplete concerning the degree to which the Earth can absorb and meet the needs of a growing population, as well as the capital and residues this population generate. There is a lot of partial information, which optimists interpret optimistically and pessimists pessimistically.
4. The continuation of current national policies in the future will not bring us closer to a desirable

state where human needs will be better met. The outcome will be an increase in the gap between rich and poor, problems with available natural resources, environmental destruction and worsened economic conditions.

5. In spite of these difficulties, current trends will not necessarily continue in the future. The world can begin a transition period in which it can direct itself toward a different future, not only quantitatively but qualitatively.

6. The exact nature of that future condition and whether it will be better or worse than the current one is not predetermined. It is the result of decisions and changes made (or not made) now.

7. When the problems are obvious to everyone, it is too late to do anything. That is why policies to change social processes should be implemented in previous stages. That is how they will have a major impact with the least consumption of resources.

8. Although technological progress is to be expected and will undoubtedly (?) be beneficial, no strictly technological change alone will lead to a better future. Social, economic and political restructuring leads to a better future in a more effective way.

9. The interdependence of people and nations through time and space is greater than what is commonly thought. Events that occur in one part of the world can have medium-term consequences that are impossible to know by intuition and probably impossible to predict, perhaps not even partially, with computer models.

10. Due to these interdependencies, simple actions that try to obtain a specific effect can be entirely counterproductive. Decisions must be made

within the broadest context of analysis of feasible areas of knowledge.

11. A focus on cooperation to attain individual or nationwide goals is frequently more beneficial in the medium-term for all the parties involved than focuses or strategies based on confrontation.

12. Many plans, programs and agreements, particularly those which are international in scope, are based on suppositions about the state of the world that are inconsistent with physical reality. Too much time and effort is spent in designing and debating policies that are, in fact, simply impossible.

For the people who have devoted time and effort in creating these world models, the common conclusions are not, after all, surprising. They have acquired an intuitive view of how the complex system in which we live functions. And the points above are no more than the key to how a complex system works.

The final conclusion of global models is quite simple: The world is a complex, interconnected, and finite system with relationships among the ecological, societal, psychological and economic scopes. We tend to act as if this were not so, as if it were divisible, separable, simple and infinite. Our persistent problems directly originate from this lack of perception. No one wants or works to produce hunger, poverty, pollution or the extinction of species. Very few people are in favour of arms use, terrorism, alcoholism or inflation. These phenomena are produced by the current system as a whole, in spite of the effort made against them. In some cases, the policies used solve problems, but many problems historically defy any solution. Perhaps we need a new perspective to solve these problems.

11. Control Questionnaire

After reading this paragraph, it is advisable that the reader answer the following questionnaire to look at what is understood and what must be reviewed before continuing.

a. Give some examples of **SYSTEM**. Remember the definition of system as a set of interrelated elements such that one element affects the behaviour of the whole set. For example: a city.

Answer:

b. Name some **ELEMENTS** of the system. For example: persons, cars, pollution, streets, etc. Incorrect elements would be: the government, the city, Barcelona, colour, asphalt, etc. They are valid as elements of the system if we can notice when the element increases or decreases, improves or worsens, etc.

Answer:

c. Name the **UNITS** of the elements. For example: Persons: number of persons, Pollution: number of particles in suspension/m3, Streets: m2.

Answer:

d. Give some examples of **CAUSAL RELATIONSHIPS** indicating whether they are positive or negative. For example: The more rainfall, the more people with umbrellas (positive).

Answer:

e. Give some examples of **LOOPS**, which are formed where there is a 'closed circuit' between two or more elements of a system. For example: The warmer you are the fewer clothes you wear (negative) and the more clothes you wear, the warmer you are (positive).

Answer:

f. Give a system example that has **JUST ONE GOAL**, indicating the goal. For example: a mower; goal: to cut grass.

Answer:

g. Give a system example that has **SEVERAL GOALS**. For example: a company, where the businessperson has the following objectives: the most profit, increasing the number of clients and increasing product quality.

Answer:

h. Give a system example that has **GOAL EROSION**. Indicate some ways to avoid contamination (in other words, erosion) of the goal by the real situation by securing it to an external element. For example: students usually have an erosion of their initial goal of getting excellent grades. In this case, an external element that can prevent this erosion is the grades of a 'rival' student.

Answer:

i. Give a system example that shows **RESISTENCE TO CHANGE**. For example: We prefer to wear our old shoes because they are more comfortable than new ones.

Answer:

j. Give a system example indicating the **LIMITING ELEMENT** that prevents an action. For example, fire does not spread because there is no more wood left; the youngster does not study because there is no more paper; the car will stop when it runs out of petrol.

Answer:

k. Give system examples and some of its **KEY ELEMENTS**. For example, the amount of salt in food is a key factor for it to be edible since, if we put in too much salt, no one will be able to eat the food.

Answer:

ANNEX

I. HISTORY AND BASIC CONCEPTS

Systems

Definitions:

a)   'A system is a set of interconnected elements (Von Bertalanffy, 1968)

b)   'A system can be defined as any set of variables selected by the observer from the whole found in a *real machine*' (Ashby, 1952)

c)   'So far, it will be enough to think of a system as a set of physical objects in a limited part of space that remain identifiable as a group throughout a significant amount of time' (Bergmann, 1957)

d)   'A set of parts put together to achieve a common goal' (Forrester)

e)   'A complex unity formed by many different facts subject to a common plan or aiming at the same purpose' (Webster's new international dictionary)

The existence of a common goal as one main feature of a system should not hide the fact that within the same system there may be conflicts of interest. For example, we might think that in a football match, each player tries to minimize effort in order to avoid exhaustion and injuries, thus taking advantage of his teammates. Overall, however, they act with the purpose of achieving the same objective, which is a final victory. Hence, the role of the team's coach consists of balancing this apparent conflict of interests so that the team works toward achieving the common goal.

Open systems

The analytic-reductionist approach aims at reducing a system to its elements in order to study them, and the systemic approach combines the elements to better understand the types of interactions between them. Bertalanffy knew that many systems, by nature, were not closed. For example, if we separate a living organism from its normal environment, it is likely to die due to an absence of oxygen, water and food. In fact, organisms are open systems that can not survive without constantly exchanging matter and energy with their environments.

In the late 1920s, Bertalanffy wrote:

'As the basic characteristic of a living being is its organization, a normal evaluation of its parts and isolated processes cannot give us a plausible explanation of vital phenomena. This evaluation does not give us information about the coordination between parts and process. Thus, the elemental task of biology should be to discover the laws of biological systems (at all levels of organization). We think that the attempts to find a basis for theoretical biology point to an essential change in the conception of the world. We will call this conception, which is considered a research method, organismic biology, and since it attempts to be explanatory, a theory for organism systems'

The systemic program was the starting point of the General System Theory, replacing the term *organism* with *organized entities* such as social groups, personalities or technological devices.

As stated by Bertalanffy (1942):

'In certain conditions, open systems are near an independent state of time called uniform state'

This uniform state is characterized by a relatively high order, represented by the existence of marked differences in the components of a system.

General System Theory

General System Theory (GST) was developed in 1940 by the biologist Ludwig von Bertalanffy. When formulated, this theory focused neither on regulation phenomena nor informational process. However, it was closer to scientific consensus than cybernetics. In fact, Bertalanffy was particularly interested in open systems.

The systemic approach is devoted to the study of the different interactions between the elements of a system and their environment. Common relationships are given in a variety of systems describing a distinctive nature. This leads to the creation of general systems. Then, it is plausible to consider a general system as a type of particular system with the same relational structure, so that any of them can function as a model of the others. This creates the need to build different theories for different systems depending on the formal context where the author is conducting his research. All the same, it is also reasonable to create a General System Theory in order to describe the specific features found in any system - not specific content but a formal mathematic theory.

A General System Theory which is successfully applicable to both real and imaginary systems should be capable of describing systems with any number of discrete or continuous variables. Thus, as Mesarovic stated, *'a system is any subset of a cartesian generalized product'*.

The importance of the interactions in a systemic approach will serve the purpose of distinguishing between input variables generated by the environment and output variables generated by the system itself. In turn, we will have to take into consideration the temporal transition in complex systems with different inner states, whether in deterministic or probabilistic processes. In cases of high systemic interest, the output in a system reacts over the input by using a feedback loop causing a non-linear process. Consequently, processes derived from regulation and balance, usually given in open systems both live and electronic, are of interest to the General System Theory.

Feedback

Cybernetics introduces the idea of circularity through the concept of feedback, breaking away from Newtonian classical science, where the effects are bound linearly.

The use of this concept might explain the evolution of social systems where two different types of feedback are found.

Cybernetics and Social Sciences

Norbert Wiener, the father of cybernetics, was a firm defender of cybernetics as a way to approach social sciences and society. Wiener was also convinced that any type of behaviour can be explained by the principles of cybernetics such as communication, control of entropy through learning by using feedback loops, etc (cf. *The Human Use of Human Beings* and *Cybernetic*s. Also, *The study of control and communication in the animal and machine*)

Cyberneticians studied nervous systems with the purpose of understanding the human condition. They concluded that observations independent from the observer are not possible. For example, when a person writes, in any language, he uses a structure in his nervous system which is the result of a history of previous interactions with languages since he was a child.

Apart from the obvious disadvantages of the researcher's subjective influence, it can sometimes act as a catalyst for change processes. Some aspects of psychology such as family therapy trace their origins back to cybernetics (Watzlawick, 1967), considering that unusual behaviors may be the result of interactions within the family. As Watzlawick states-

> *We basically affirm that interpersonal systems - groups of strangers, couples, families, psychotherapeutic or even international relations - can be understood as feedback circuits, since a person's behavior affects and is affected by others.*

In the following pages an approach which can be used to interpret reality will be described. The *correct* or *the best* way to observe reality does not exist, given that it is impossible to define a specific path as the best or most accurate.

Cybernetics

The word cybernetics originates from the Greek word 'Kybernetes' which appeared for the first time in the writings of Plato and later, in the 19th century, André Marie Ampere used the word to refer to different forms of government.

By 1943, a group of scientists, led by the mathematician Nobert Wiener, had recognized the need to choose an appropriate word to describe this array of theories and concepts. In 1947, the group adopted the word 'Cybernetics' which gained popularity in Weiner's book, *'Cybernetics, or the Study of Control and Communication in the Animal and Machine'* (1948). Since then, the discipline has received an increasing degree of interest. Cybernetics has expanded both as an interdisciplinary science which aims to study control or self-control (Wiener), and also as science which seeks the efficacy of action (Couffignal).

State of Space

Cybernetics addresses the difference between the presence and absence of specific features, also referred to as dimensions or attributes. For example, a system called billiard ball may have features such as a particular colour, weight, position or speed. The presence or absence of each feature can be represented with Boolean variables that take on two values: 'yes' when the system has the feature, or 'no' when the system does not have the feature.

The binary representation can be generalized to a subsequent feature with multiple, discrete or continuous values. The set of all possible states of a system is referred to as State of Space. An essential component of cybernetic modeling is the quantitative measure of State of Space's dimension, or the number of different states. This measure is called variety. The variety is defined by the number of elements in a Space of State: $V=\log([S])$

Entropy

The concept of entropy was introduced by Clausius during the 19th century. It was intended to act as a measure of disorder in gas molecules in order to balance thermodynamic accountability.

Statistically, disorder exists because of the number of distinct states a system is capable of adopting. If one compares two systems, system A will have more disorder than system B whenever the number of distinct states found in A is higher than in B.

In a closed system, entropy increases according to Clausius equation-

$$dS>0$$

whereas in an open system the total change of entropy can be expressed as specified by Prigogine-

$$dS=dSi+dSe$$

where dSe denotes the change of entropy due to importation that can be either negative or positive, and dSi represents the production of entropy due to irreversible processes, which is always positive.

II. FREQUENTLY ASKED QUESTIONS

What is the difference between a process of Addiction and Shifting the burden?

This question is interesting because of the significance it implies. In both situations the system manages to equalise the Real State with the Desired State with external help.

We talk about Addiction when an object - a thing - intervenes and by Shifting the burden when another system intervenes with its own objectives.

The consequences of this detail are important because the object of an Addiction would never leave us and therefore we don't have to expect any change if we don't want one. On the contrary, the system that supports our charge today can decide tomorrow to stop supporting us and provoke a crisis.

For example, we can be addicted to tobacco and in this case, if we manage to reduce our stress with this practice we can be sure that we will always be able to do it as long as we don't decide to give up smoking. On the contrary, if we have passed the charge of our low incomes to our father, it's possible that one unexpected day the subject of our charge decides that he has already been patient with us and he stops helping us.

What is the difference between the limiting factors and the key factors?

The key factors are elements of the system that are very sensitive. They are always the same. Anyone would be sensitive if somebody put a finger in their eye and would probably react with violence. But in reality we have two eyes and it would not be life threatening to lose one of them.

Each system has its own key factors and in order to discover them we need to invest a certain amount of time and effort. It is important to understand these key factors if we want to manipulate the system without altering any factors that could provoke a violent response. On the other hand, we need to try to take advantage of the factors that produce a positive reaction in the system. It is very important to remember that in general they are hidden and that they are always the same.

The limiting factors, on the other hand, are usually very visible and they usually change with time. They are the elements which will condition the state of the system now or in the immediate future. However, tomorrow they could be other different elements. For example, I'm hungry so I don't work - I go to have something to eat. Once I have eaten, the limiting factor is that I don't have paper, so I go to get paper. When I have paper, I don't have any ideas. In other words the limiting factors are constantly changing.

Which temporal horizon should we define?

This is an essential aspect that requires special attention in each model. We need to be generous in the definition of the time limit of the simulation. Restrictions from the point of view of the hardware or the software don't exist. The existing software executes simulations in just a few seconds.

We need to avoid focusing too much on the temporal horizon that the client or user suggests. Sometimes, certain phenomena can manifest in the model shortly after the temporal horizon chosen, that can also show themselves shortly before inside the horizon that we have chosen.

A wide temporal horizon allows us to have the security that certain phenomena are really what they seem to be, in the way that a system with stable oscillations doesn't start to grow - and are therefore unstable - after a certain determined period.

III. TRAINING COURSES

This book attempts to offer the necessary formation in order to gain a detailed knowledge of the subject, both from theoretical and practical points of view. Nevertheless, the values of a course can not be underestimated. A course will provide immediate answers to questions of particular interest for the student.

The creation of simulation models in general and System Dynamics in particular can be found integrated into the study plans of some universities. Additionally, there are many centres in the world that offer postgraduate or doctorate courses to specialise in this subject. For years, Internet has shown itself to be an excellent source of training material as the relationship between teacher and student can be very personal. Also, models can be exchanged quickly and easily.

In general, the courses that are offered have three different areas of interest- environmental, business, and social.

The courses in the environmental area are based on the realisation of ecological and biological studies, the management of natural resources, studies of environmental impact, environmental consulting and management of the industrial environment, the educational environment and diagnostic environment and in general, in the studies based on the relationship between humans and the environment such as urban or regional planning.

In the business area, these models can be applied in the realisation of planning strategy studies, project management, sector studies and in general, all those areas where traditional techniques of optimisation are not applicable because of the complexity or the existence of relevant qualitative aspects.

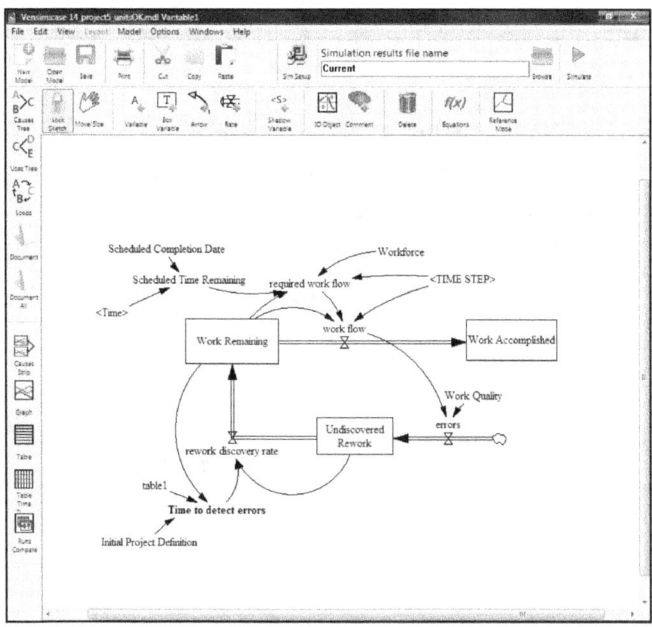

Lastly, the most innovative area where these systems can be applied is in the social area, especially in the realisation of studies in psychology, psychiatry, sociology, family help, direction, organisation and administration of companies, local and regional planning, human resource management, foundations and NGOs, consultancies, and in medicine as a support to the professors.

A good training course should include the theoretical and practical aspects of how to manage the software. This is important because in many cases the student is only interested in receiving training on the practical aspects, in particular the use of the software. The theoretical training is an excellent opportunity to become familiar with the possible applications of System Dynamics, its positive aspects and its limitations.

Sincerely, if you want to know what System Dynamics is and how it works, it is advisable to take a course. If this is not possible, it is necessary to be aware of the limitations and reinforce where possible with some theoretical reading.

More information about the System Dynamic courses in: **http://atc-innova.com**

IV. SOFTWARE

The appropriate software to achieve simulation models applying System Dynamics has evolved in recent years in two ways. One way is to become more user-friendly thanks to the use of Windows. The way has been the continual increase of its performance features.

Today in the market, there are a few big software companies and other smaller ones that can create theses type of models. It is necessary to enter into subtle details to achieve an objective evaluation of the advantages and the limitations of some software in relation to others. Usually, many of the aspects that can lead us to choose one particular software over another are often subjective.

In this book, I don't try to offer a comparative evaluation of the various software marks in the market even though I don't intend to leave the reader without any guidance about the software to use. If the reader doesn't have any software simulation for these types of models, my advice is to obtain by internet **Vensim PLE®** which is free for educative or personal uses and has important features that other software doesn't offer in their free versions, for example it is possible to save the model that you have created. As an additional reference about the quality of the software Vensim, it is actually the software used for teaching and investigation in the Sloan School of Management in MIT (Massachusetts Institute of Technology.) To obtain this software you can go to the web http://atc-innova.com. This company usually launches new versions of the software every year with improvements and additional features therefore, it is possible that the CD of this book may contain a version that is not fully updated.

The level of the features of the software for educative use allows the creation of models that during the 80s required the most powerful computers. Therefore, it will be very difficult for a person creating their first models to find the limits in this software.

Other software marks widely used are Stella and ithink from the company HPS. Information can be found on their web: http://www.hps-inc.com.

The software ithink has specific applications in the economy and business management. It has some very powerful graphic features that are user-friendly.

Stella was launched in 1985 for the Mac equipment and was a real revolution that allowed the user to discard the old program Dynamo as it offered a visualisation of the relation between elements and functions that were non-lineal. Its principal destination at the present can be found in the applications in scientific investigation and teaching.

The applications of Powersim can be found in the business area. They are mainly financial models, management of clients, analysis of production, human resources and the development of new products.

V. BIBLIOGRAPHY

Papeback   ISBN: 978-1718096264
Kindle ebook   ASIN: B07G5KQ7DG

Ackoff, R.L. (1974) *Redesigning the Future*. New York NY. John Wiley.

Allison, J.A. et al. (1994) *Uses of modeling in science and society*. Ethics in Modeling. White Plains. Pergamon. New York.

Anderson, Virginia. (1997) *Systems Thinking Basics: From Concepts to Causal Loops*. Waltham MA. Pegasus Communications.

Checkland, P. (1999). *Systems Thinking, Systems Practice: A 30-Year Retrospective*. John Wiley and Sons. New York

Drucker, P. (1995). *Managing in a Time of Great Change*. New York NY: Dutton.

Forrester, Jay W. (1961) *Industrial Dynamics*. Waltham MA. Pegasus Communications.

Forrester, J. W. (1969). *Urban Dynamics*. Norwalk, CT: Productivity Press.

Forrester, J. W. (1971). *Principles of Systems*. Norwalk, CT: Productivity Press.

Forrester, J. W. (1973). *World Dynamics*. Norwalk, CT: Productivity Press.

Forrester, J. W. (1975). *Collected Papers of Jay W. Forrester*. Norwalk, CT: Productivity Press.

Goodman, M. R. (1974). *Study Notes in System Dynamics*. The MIT Press.

Haines, S. G. (1998). *The Managers Pocket Guide to Systems Thinking and Learning*. Amherst, MA: HRD Press

Haines, Stephen G. (1999) *The Systems Thinking Approach to Strategic Planning and Management.* Amherst MA HRD Press.

Haines, S. G. (2000). *The Complete Guide to Systems Thinking and Learning.* Amherst, MA: HRD Press.

Kerzner, Harold. (2004) *Project Management Best Practices: Achieving Global Excellence.* Hoboken, NJ. John Wiley & Sons.

Kim, D. (1994*). Systems Thinking Tools: A User's Reference Guide.* Cambridge, MA: Pegasus Communications.

Kim, Daniel H. (1998*) Introduction to Systems Thinking.* Waltham MA. Pegasus Communications.

Livsey, Rachel C. (1999) *The Courage to Teach, A Guide for Reflection and Renewal.* Hoboken, NJ. John Wiley & Sons.

Meadows, Dennis L. (1970*) Dynamics of Commodity Production Cycles.* Waltham, MA. Pegasus Communications.

Meadows, Dennis L. (1974) *Dynamics of Growth in a Finite World.* Waltham, MA. Pegasus Communications.

Meadows, Donella an Dennis et al. (2004) *Limits to Growth: The 30-Year Update.* White River Jct., VT Chelsea Green Publishing

Mesarovic, M. (1967). *Views on General Systems Theory.* NY: John Wiley and Sons, Inc.

Morecroft, John D.W. and Sterman, John D. (1994) *Modeling for Learning Organizations.* New York, NY. Productivity Press.

Miller, G. (1978). *The Need for a General Theory of Living Systems.* Mc Graw Hill. New York.

Mulgan, G. (1997). *Connexity: How to live in a Connected World.* MA: Harvard Business School Press.

Naisbitt, J. (1994). *Global Paradox: The bigger the World Economy, the More Powerful Its Smallest Players.* NY: William Morrow.

Naisbitt, J. - Aburdene, P. (1990). *Megatrends 2000: Ten New Directions For The 1990's.* NY: William Morrow and Company, Inc.

Richardson, George P. and Pugh III, Alexander L. (1981) *Introduction to System Dynamics Modeling.* Waltham, MA. Pegasus Communications.

Richardson, G. P. (1991). *Feedback Thought in Social Science and Systems Theory.* Philadelphia, PA: University of Pennsylvania Press.

Roberts, E. B. (1981). *Managerial Applications of System Dynamics.* Norwalk, CT: Productivity Press.

Roberts, N. (1983) *Introduction to Computer Simulation.* NY. Addison-Wesley Publishing Company.

Peter Senge (1980) *A system dynamic approach to investment-function specification and testing.* Socio-Economic Planning Sciences 14.

Peter Senge (1990) *The Fifth Discipline.* The Art and Practice of Learning Organizations. New Yor NY. Doubleday.

Peter Senge and John Sterman (1992*) System thinking and organizational learning: Acting locally and thinking globally in the organization of the future.* European Journal of Operational Research 59.

Peter Senge (1994) *The Fifth Discipline Fieldbook.* New Yor NY. Doubleday.

John Sterman (2000). *Business Dynamics: Systems Thinking and Modeling for a Complex World.* NY: McGraw-Hill Higher Education.

John Sterman (1994) *Modeling for learning organizations.* Edited with John D. W. Morecroft

John Sterman (2000) *Business Dynamics: Systems thinking and modeling for a complex world.* McGraw Hill.

John Sterman (2005) *Operational and Behavioral Causes of Supply Chain Instability.*

Van Kavelaar, Eileen K. (1997) *Conducting Training Workshops: A Crash Course for Beginners* . Hoboken, NJ. John Wiley & Sons.

Von Bertalanffy, L. (1998). *General Systems Theory: Foundations, Development, Applications.* NY: George Braziller, Inc.

Weiner, N. (1965). *Cybernetics or Control and Communication in the Animal.* Cambridge, MA: MIT Press.

VI. ACKNOWLEDGEMENTS

Without a doubt this book is indebted to those who were my teachers for having signalled the importance of this discipline as an instrument of analysis. I am also indebted to some friends, José Alfonso Delgado who convinced me of the need to have books available on this subject and encouraged me to dedicate the effort needed to give form to the book.

Secondly, this book is the product of many years of teaching and therefore **my greatest thanks are to all my students** who, with their continuous questions, have made me reflect on the theoretical concepts and make the examples clearer and simpler.

Those who have collaborated in this book:

- Mario Guido Pérez (Chemical Engineer, Argentina) author of the models of the Chemical Reactor, Ingestion of Toxics, Golden Number and Butterfly Effect.

- Claudio M. Enrique (UNL, Santa Fe, Argentina) author of the model Study of the Oscillatory Movements.

- Gustavo Adolfo Juarez (University de Catamarca, Argentina) for his collaboration in the model Development of an Epidemic.

- José Ignacio Fernandez Mendez (UNAM, Mexico) and Michael Frenchman (Consultant, USA) in the model of the fishery of shrimp in Campeche.

- Josep Maria Banyeres (Engineer, Spain) in the Barays of Angkor model.

- Mohamed Nemiche (Doctor in Physical Sciences; Morocco) in the part of history and basic concepts.

- Antoni Lacasa Ruiz (Artist, Spain) who has offered his experience in the drawings that illustrate and make more agreeable the text.

... and all the translators of the original book, specially to the reviewer Terry Walker.

Now you can continue with this book ...

A book for Vensim PLE PLUS users

ISBN 979-8655650183

Includes examples of the most common mistakes in building a simulation model.

ISBN 979-8662618657

Collection of books

Selected papers on System Dynamics

Modeling Agriculture and Food Production
ISBN: 9781686984570

Modeling and Simulating Business Dynamics
ISBN: 9781686997556

Paper 1. New Technologies and Employment
Paper 2. Dynamic Balanced Scorecard
Paper 3. The Procurement Process
Paper 4. Scenario Planning Workshop
Paper 5. Risk Analysis Methods
Paper 6. Stereotypes in Socio-Economic Systems
Paper 7. Enterprise Resource Planning
Paper 8. Marketing Research
Paper 9. Group Model Building
Paper 10. Business Dynamics Simulator
Paper 11. Strategic Decision Support
Paper 12. Rare Earth Elements
Paper 13. Building a Learning Lab

Modeling in Ecology and the Environment
ISBN: 9781687000323

Modeling the Economy: Money and Finances
ISBN: 9781687003133

Modeling and Simulation in Energy Management
ISBN: 9781687004932

Paper 1. Modeling and Simulating Energy Policies
Paper 2. Water-Energy-Food Nexus
Paper 3. Environmental-Social Pressures in Mining
Paper 4. Impacts of Electric Vehicle Diffusion
Paper 5. Forecasting Electricity Demand Market
Paper 6. Rare Earths Production Forecasting
Paper 7. Risk Analysis of Offshore Fire
Paper 8. Scenario Planning Implementation
Paper 9. Energy and Environmental Protection
Paper 10. Simulating Petroleum Peak Curve
Paper 11. Participative Group Model Building
Paper 12. Green Growth and Ecotax
Paper 13. Enterprise Resource Planning Implementat.

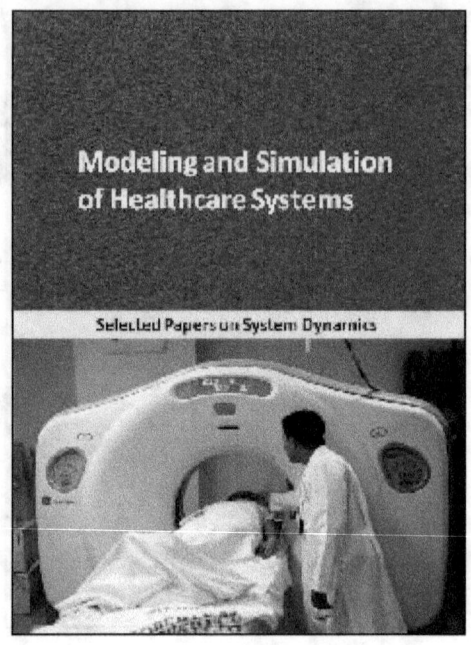

Modeling and Simulation of Healthcare Systems
ISBN: 9781687006745

Modeling the Housing and Urban Dynamics
ISBN: 9781687008367

Modeling Supply Chains and Industrial Dynamics
ISBN: 9781687009975

Modeling Labor, Human Resources, and Social
Dynamics
ISBN: 9781687015389

Modeling Sustainable Development
ISBN: 9781700341600

JOIN NOW!

Vensim Online Courses

http://vensim.com/online-courses/

www.ingramcontent.com/pod-product-compliance
Lightning Source LLC
Chambersburg PA
CBHW071214220526
45468CB00002B/594